I0821670

Nature's Mysteries

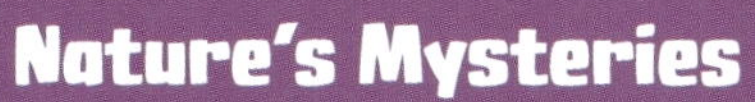

AUSTRALIA'S PINK LAKES

by Patricia Hutchison

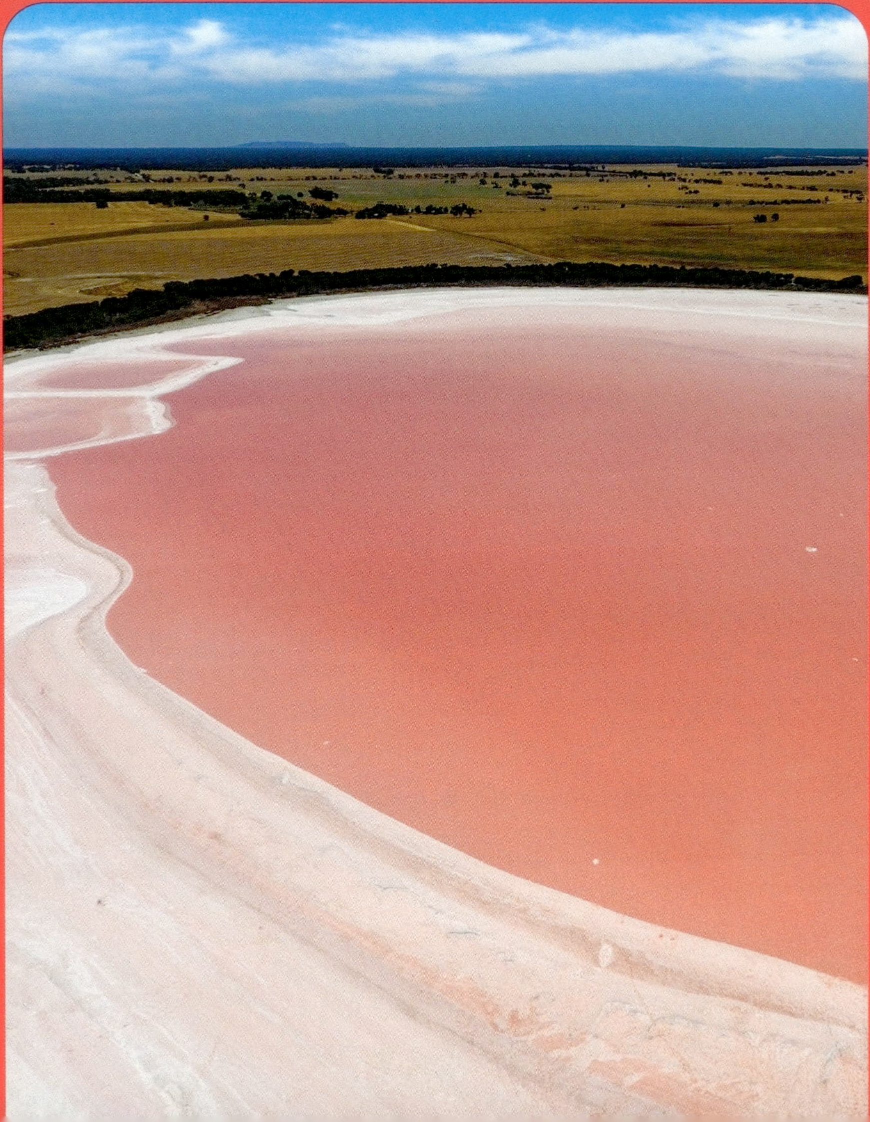

abdobooks.com

Published by Pop!, a division of ABDO, PO Box 398166, Minneapolis, Minnesota 55439.

Printed in the United States of America, North Mankato, Minnesota.

102020
012021

Cover Photo: iStockphoto
Interior Photos: iStockphoto, 1, 29; Shutterstock Images, 5, 6, 8, 9, 13, 21, 27, 28; Stuart Forster/robertharding/Alamy, 11; Terra incognita/Alamy, 14; Piter Lenk/Alamy, 15; Alexander Kondakov/Alamy, 17; Frank Fox/Science Source, 18–19; Red Line Editorial, 22–23; Peter Bischoff/PB Archive/Getty Images, 25

Editor: Alyssa Krekelberg
Series Designers: Candice Keimig, Victoria Bates, and Laura Graphenteen

Library of Congress Control Number: 2020940294

Publisher's Cataloging-in-Publication Data

Names: Hutchison, Patricia, author.

Title: Australia's pink lakes / by Patricia Hutchison

Description: Minneapolis, Minnesota : POP!, 2021 | Series: Nature's mysteries | Includes online resources and index

Identifiers: ISBN 9781532169168 (lib. bdg.) | ISBN 9781532169526 (ebook)

Subjects: LCSH: Lakes--Juvenile literature. | Salt lakes--Juvenile literature. | Microalgae--Juvenile literature. | Curiosities and wonders--Juvenile literature. | Mystery--Juvenile literature. | Geography--Juvenile literature.

Classification: DDC 910.02--dc23

WELCOME TO DiscoverRoo!

Pop open this book and you'll find QR codes loaded with information, so you can learn even more!

Scan this code* and others like it while you read, or visit the website below to make this book pop!

popbooksonline.com/pink-lakes

*Scanning QR codes requires a web-enabled smart device with a QR code reader app and a camera.

TABLE OF CONTENTS

CHAPTER 1

PINK WATER

Tourists fly over Australia in a helicopter. They look down at a green island off the country's southwestern coast. A small lake sits among the trees. It's unlike any lake they have seen before. The water is bright pink.

WATCH A VIDEO HERE!

Pink lakes add even more color to the already beautiful Australian landscape.

The north park of Lake Hillier is surrounded by paperbark and eucalyptus trees.

Lake Hillier is on Australia's Middle Island. A strip of sand and trees separate it from the ocean. The shallow lake is only 1,969 feet (600 m) long. But its striking pink color catches people's attention.

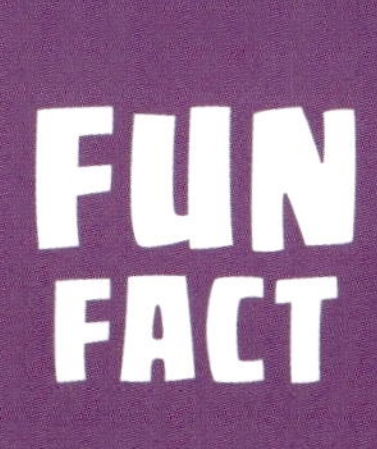

People didn't discover Lake Hillier until 1802. That year, a man climbed to the island's highest point and looked down. He quickly noticed the lake's strange color.

Tourists can reach some pink lakes by car. To see others, people must fly or ride a boat.

Lake Hillier is a favorite stop for tourists. But it is not the only pink lake. Australia has many pink lakes scattered throughout the country.

WHERE ARE SOME OF AUSTRALIA'S PINK LAKES?

1. Hutt Lagoon
2. Pink Lake of Quairading
3. Lake Hillier
4. Lake MacDonnell
5. Lake Eyre
6. Lake Bumbunga
7. Lake Albert
8. Murray-Sunset National Park has four pink lakes: Lake Becking, Lake Crosbie, Lake Kenyon, and Lake Hardy
9. Westgate Park

CHAPTER 2

CHANGING COLORS

Lake Hillier stays bright pink all year round. But other pink lakes and bodies of water in Australia change colors. For example, Hutt Lagoon can look pink, red, or purple. Its color depends on the time of day or season.

LEARN MORE HERE!

Hutt Lagoon is approximately 27 square miles (70 sq km) in area.

Lake Bumbunga is easy for people to travel to. Many people visit the lake to take photographs of it. Some people like wading in the colorful water. The water appears as a mixture of different shades of pink, white, and blue. Its color depends on how much salt is in the water at the time.

Pink Lake of Quairading (KWEE-ruh-ding) has a road running through the middle. Each side is a different shade of pink.

It takes less than two hours to reach Lake Bumbunga from the city of Adelaide.

Visitors will find four pink lakes in Murray-Sunset National Park in Victoria, Australia. Their names are Lake Becking,

People can camp in Murray-Sunset National Park.

Lake Crosbie can take on a whitish-pink coloring.

Lake Crosbie, Lake Hardy, and Lake Kenyon. These lakes are brightest on cloudy days. They can change from brilliant pink to gleaming white.

CHAPTER 3

WHY ARE SOME LAKES PINK?

Scientists pondered the mystery of pink lakes for years. Researchers think they have finally solved it. One reason lakes are pink is that they have high **concentrations** of salt. This happens especially when the climate is dry.

LEARN MORE HERE!

People can find hard salt on the shores of pink lakes.

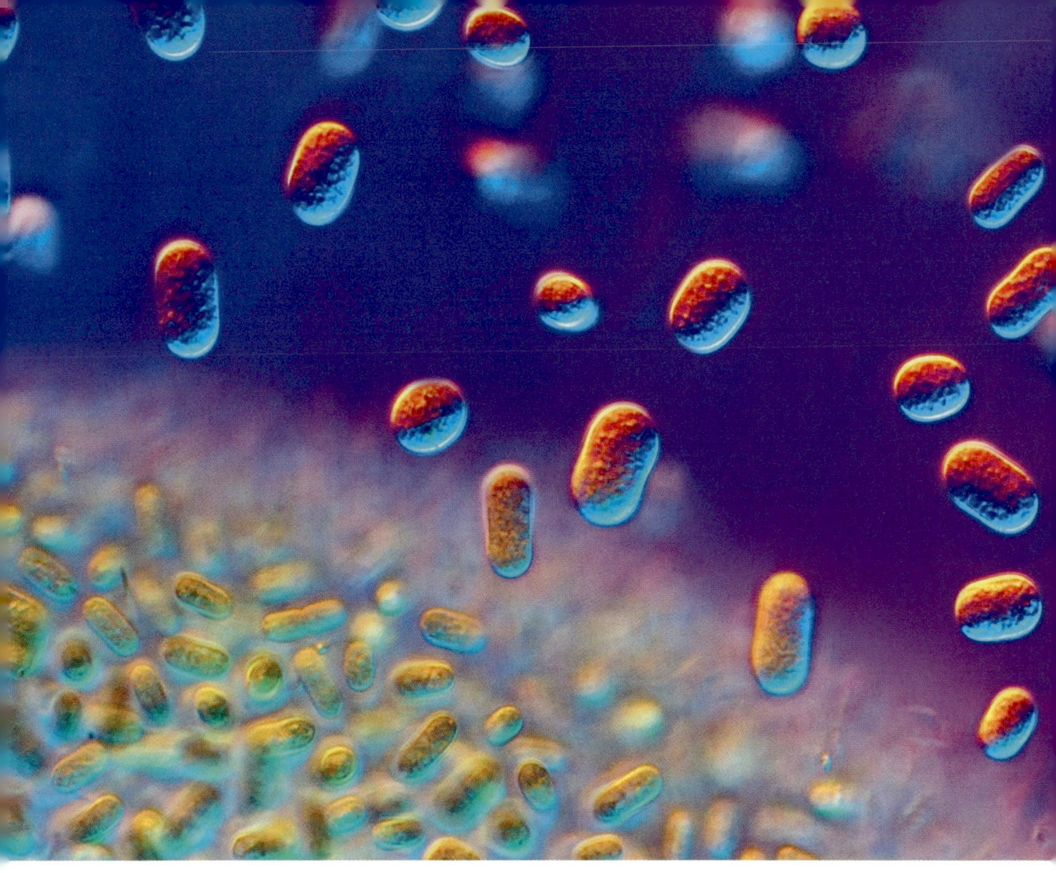

Scientists also figured out that the salty lakes are filled with a certain type of **algae** that have red **pigment**. These algae need salt water to survive.

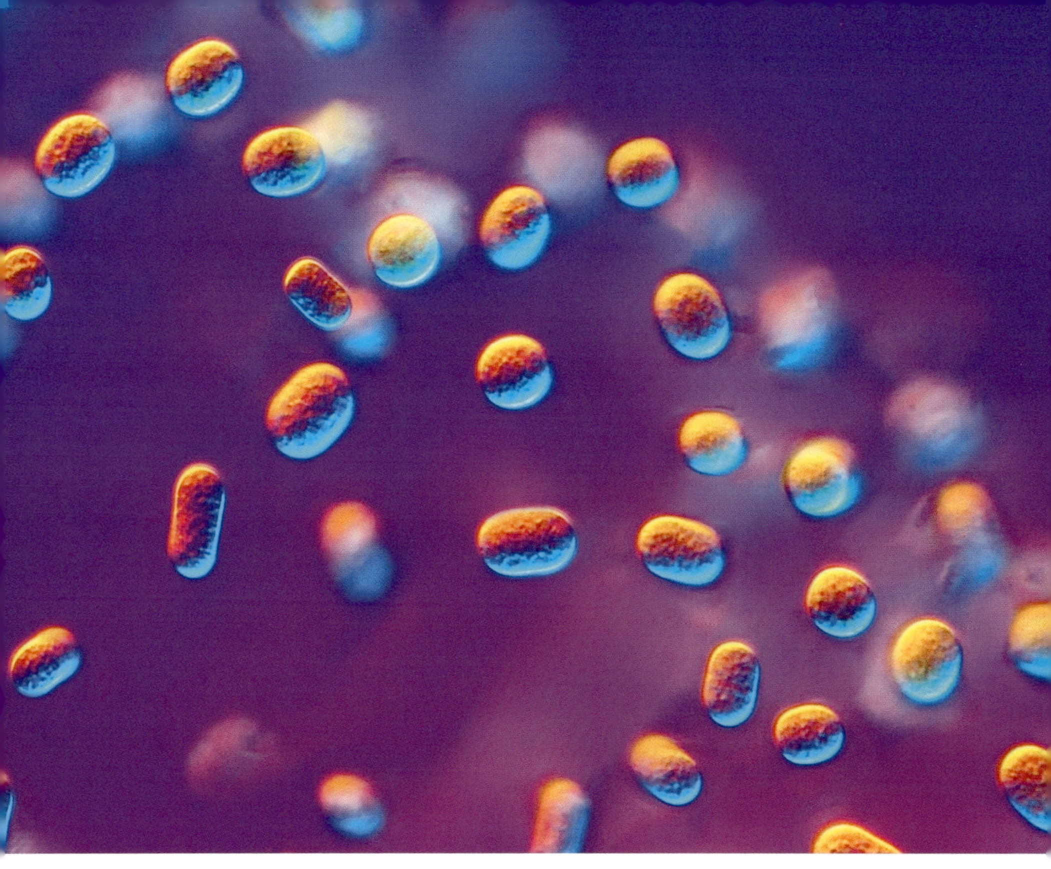

The algae in pink lakes are so small that people need a microscope to see them.

FUN FACT

People can swim in some pink lakes, but they should not try to drink the water. It is too salty!

MINING SALT

In the early 1900s, people **harvested** salt from pink lakes in what is now Murray-Sunset National Park. They used shovels, wheelbarrows, and other tools to pick apart the salt crust. Then, camel teams carried the salt to the railway. Mining stopped when the area became a national park in 1979.

In 2015, scientists found a type of **bacteria** growing in the salty crust at the bottom of Lake Hillier. The bacteria have a pink pigment. Together, the bacteria and the algae make the salty lakes look pink.

Most lakes have fresh water, meaning they don't have any salt. That's another reason why pink lakes are so unusual.

POPULAR PINK LAKES AROUND THE WORLD

Pink lakes are found across the globe. All of them have salt. Some are salty because of the dry climate. In others, the salty water comes from being close to the sea.

CHAPTER 4

PINK LAKE FADES

A lake's pink color can sometimes change. For example, Pink Lake is in a town called Esperance, Australia. The lake used to be a beautiful pink color. But human activities started to affect the lake. Over many years, the color faded.

COMPLETE AN ACTIVITY HERE!

Scientists are trying to find ways to make Pink Lake pink again.

Years ago, people built a highway and railway near the town. Suddenly, salt water from other lakes couldn't get into Pink Lake. The **algae** could not survive to make the red **pigment**. The lake's color faded. Now, the town wants to change Pink Lake's name. That's because when people visit the lake, they are disappointed by its white color.

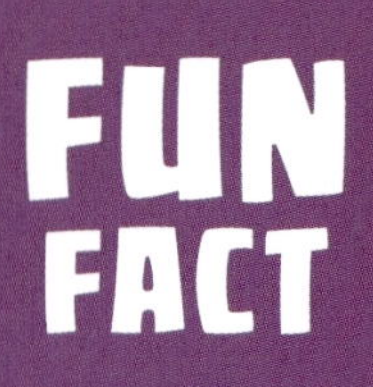

People mined the salt in Pink Lake from 1896 to 2007. This took away some of Pink Lake's color.

People still mine salt in some pink lakes. They take salt from the lake and use shovels to toss it into containers.

Local governments ask tourists to leave the areas around pink lakes undisturbed.

People often change the environment and ruin natural wonders. Australia is trying to protect its other pink lakes. Sometimes, visitors leave trash behind. This can hurt the lakes.

State governments have limited activities such as camping near the pink lakes. This helps reduce the number of people at the lakes. They hope the lakes will keep their stunning pink color for many years.

Rottnest Island in Western Australia also has a bright pink lake.

MAKING CONNECTIONS

TEXT-TO-SELF

What is the most exciting thing in nature that you have ever seen? What was special about it?

TEXT-TO-TEXT

Have you read other books about natural wonders? How are they similar to or different from Australia's pink lakes?

TEXT-TO-WORLD

Australians are trying to protect their pink lakes. What are some natural wonders in the area where you live? What can people do to protect them?

GLOSSARY

algae – plants that live in ponds or lakes and do not have stems, flowers, or roots.

bacteria – tiny living things.

concentration – the amount of an ingredient in a mixture.

harvest – to gather or pick something.

pigment – a natural substance that gives color to other materials.

tourist – someone who visits a place for enjoyment.

INDEX

ONLINE RESOURCES

popbooksonline.com

Scan this code* and others like it while you read, or visit the website below to make this book pop!

popbooksonline.com/pink-lakes

*Scanning QR codes requires a web-enabled smart device with a QR code reader app and a camera.